OS2U Kingfishe

by Al Adcock
Color by Don Greer/Perry Manley
Illustrated by Joe Sewell

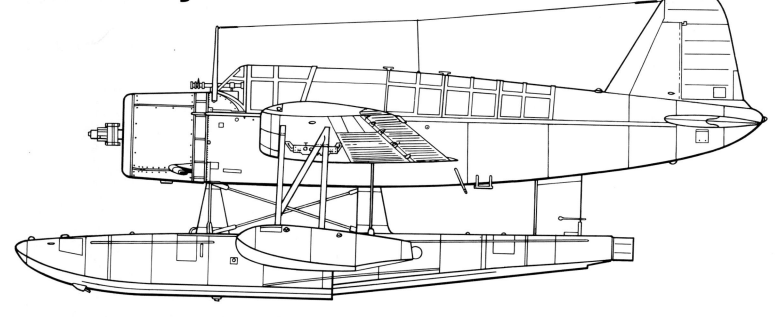

Aircraft Number 119
squadron/signal publications

An OS2U-1 Kingfisher takes off after picking up a downed Navy pilot. Kingfishers served in the rescue, gunfire spotting, scouting and anti-submarine roles throughout the Second World War.

COPYRIGHT © 1991 SQUADRON/SIGNAL PUBLICATIONS, INC.
1115 CROWLEY DRIVE   CARROLLTON, TEXAS 75011-5010
All rights reserved. No part of this publication may be reproduced, stored in a retrieval system or transmitted in any form by any means electrical, mechanical or otherwise, without written permission of the publisher.

ISBN 0-89747-270-5

If you have any photographs of the aircraft, armor, soldiers or ships of any nation, particularly wartime snapshots, why not share them with us and help make Squadron/Signal's books all the more interesting and complete in the future. Any photograph sent to us will be copied and the original returned. The donor will be fully credited for any photos used. Please send them to:

Squadron/Signal Publications, Inc.
1115 Crowley Drive.
Carrollton, TX 75011-5010.

Acknowledgements and Photo Credits

LTV
United Technologies
National Archives
Bureau of Ships
Fred C. Dickey, Jr.
USAF Armament Museum
Northrop
EDO Corporation
U.S. Navy
Nicholas J. Waters III
LCDR R.E. Geale, MBE, RAN (Ret)
Jim Presley
John M. Elliott
D. W. Gardner

Jim Croslin
Anne Millbrooke
Paul White
John Riley USN (Ret)
Aviation Photo Exchange
Howard K. Corns
Ira E. Chart
J. J. Frey
Bill Johnson
Museo De Chile
Paul J. McDaniel
Navy History Center
Bob Krenkel

The pilot and crewman of this beached OS2U-3 Kingfisher discuss their mission as ground crewmen go over the aircraft. The recently painted out Red disc on the underwing insignia puts the time frame at June of 1942. The OS2U was used for everything from training to anti-submarine patrols. (Edo Corp.)

Introduction

Before the advent of radar, the scout/observation aircraft served as the eyes of the fleet. Scouting hundreds of miles either from a shore base or from a ship these aircraft were used to observe submarine and/or surface ship movements, act as gunfire spotters for surface ships, and act as air-sea rescue platforms.

The Chance Vought Aircraft Company came into existence during 1917, producing trainer aircraft for the U.S. Army. Later the Lewis and Vought VE-7 series of aircraft was procured by the Navy. The VE-7 series developed into the VE-9 with one model, the VE-9H, being produced as a single float observation aircraft (four were built for the Navy). During the 1920s Vought brought out the UO-1, a variant of the VE-9. The aircraft was powered by a 220 hp Wright J-5 engine and a total of 141 UO-1s were built. These became the first service aircraft to operate from the deck of an aircraft carrier and were also the first aircraft to operate from a battleship catapult. The UO-1 was followed by the Vought O2U-1, the first in a long line of Corsair biplanes. Vought continued to produce the Corsair biplane series, up to the O3U-6 introduced during 1935. The O3U-6 would be the last biplane produced by Vought for the Navy.

During 1937, the U.S. Navy asked all interested aircraft manufacturing companies to submit bids for a new observation scout aircraft that could be either land or water based. Three companies submitted aircraft to meet the specification which called for a two seat observation aircraft that would be small enough to operate without folding wings from a battleship.

The entry from Stearman was the Model 85, a tandem two seat biplane of mixed metal and fabric construction with interchangeable wheel and float landing gear. Given the designation XOSS-1, a single prototype (BuNo 1052) was ordered on 6 May 1937 and flown sixteen months later during September of 1938. The XOSS-1 was powered by a 600 hp Pratt and Whitney R-1340-36 radial engine and had a top speed of 162 mph at 6,000 feet. The wing span was thirty-six feet (as called for in the specification) and the length was thirty-four feet six inches when float equipped. The XOSS-1 featured full length flaps on the upper wing which gave it a landing speed of 57 mph.

Another entry was the XOSN-1 produced by the Naval Aircraft Factory, Philadelphia Navy Yard, Pennsylvania. This aircraft was also a tandem two seat biplane of mixed metal and fabric construction. The XOSN-1 (BuNo 0385) featured automatic leading edge wing slats and I struts between the wings (which was a departure from the customary N struts normally found on biplanes). The XOSN-1 was powered by the same engine as the Stearman entry and had a top speed of 160 mph at 6,000 feet. Landing speed, thanks to the automatic slats, was 54 mph when float equipped. Wing span and overall length was the same as the XOSS-1. The Navy found nothing new in these two entries and began looking very closely at the entry from Vought-Sikorsky.

Vought's entry, the XOS2U-1 (Model VS310) was a radical departure from the other entries. The XOS2U-1 prototype was ordered by the Bureau of Aeronautics on 19 March 1937 under contract number 53737. The prototype was a two place mid-wing monoplane of mixed construction. The XOS2U-1 featured a number of innovations: it would be the first production aircraft to employ spot welding and the fuselage was of all aluminum alloy monocoque construction with only the rear portions of the wing and all moveable control surfaces being fabric covered (with the exception of the wing spoilers).

The XOS2U-1 was powered by a 450 hp Pratt & Whitney R-985-4 Wasp nine cylinder air cooled radial engine driving a Hamilton-Standard two blade controllable pitch propeller. Designed by Rex B. Beisel, the XOS2U-1 was the first monoplane ever catapulted from a shipboard catapult. The prototype (BuNo 0951) made its first flight on 1 March 1938 as a land plane. The aircraft was later converted to a float plane with the addition of a centerline float and two wingtip floats. It was flown in this configuration on 19 May 1938.

The XOS2U-1's first flight as a landplane was a full three months ahead of the entries from the NAF and Stearman. Early test flights revealed that the XOS2U-1 had a top speed (on floats) of 171 mph at 5,000 feet and a landing speed of 55 mph. The low landing speed was accomplished by the use of full span spring loaded flaps, a very low wing loading

The Stearman XOSS-1 (Model 85) was delivered to the Navy during August of 1938. The aircraft was ordered as a backup against possible failure of the Vought XOS2U-1 prototype. (Via Bob Krenkel)

The experimental department at Vought-Sikorsky was responsible for the design and construction of the wooden XOS2U-1 mock-up. Vought had a long history of producing observation/scouting aircraft for Navy and this was their first monoplane design accepted by the Navy. (National Archives)

and a large wing area of 262 square feet. Lateral or aileron control, when the ailerons were dropped along with the flaps, was accomplished by means of an upper wing spoiler. The XOS2U-1 weighed 3,213 pounds, had a wing span of thirty-five feet eleven inches and a length of thirty-three feet seven inches. The prototype featured a rounded leading edge of the rear observer's greenhouse.

When presented to the Navy at NAS Anacostia, Washington, D.C. for evaluation on 2 August 1938, the XOS2U-1 was painted in the then standard overall Aluminum Dope with Chrome Yellow upper wing surfaces. When it was sent to the USS WEST VIRGINIA for catapult trials, the XOS2U-1 was brightly painted with a Dark Blue fuselage, wing and horizontal tail, Aluminum floats and rudder. The aircraft carried this scheme during its fleet acceptance trials. These trials proved very successful and the Navy placed an order with Vought for fifty-four production aircraft under the designation OS2U-1 (contract number 66259).

During 1939, the Chance-Vought Aircraft Division of East Hartford Connecticut, and Sikorsky Aircraft (a division of United Aircraft), in Stratford, merged their aircraft manufacturing facilities at Stratford, enlarging the plant.

For sea trials and acceptance flights at NAS Anacostia, Washington, D.C., the fuselage wings and stabilizers of the prototype XOS2U-1 were painted in a Dark Navy Blue. The XOS2U-1 was catapult tested aboard the USS WEST VIRGINIA (BB-48) during early 1939 and was officially accepted on 24 April 1939. (LTV)

Vought test pilot Paul S. Baker taxies the prototype out toward Long Island Sound on a test flight. The XOS2U-1 first flew as a seaplane on 19 May 1938. Tests by Vought and the Navy were successful and the Navy ordered a total of fifty-four aircraft on 22 May 1939. The XOS2U-1 remained active as a test aircraft until 20 May 1944. (LTV)

The prototype XOS2U-1 (BuNo 0951) made its first flight as a land plane on 1 March 1938, from the company field at Stratford, Connecticut. The prototype was powered by a 450 hp Pratt and Whitney R-985-4 air cooled radial engine giving it a top speed of 175 mph. (LTV)

Development

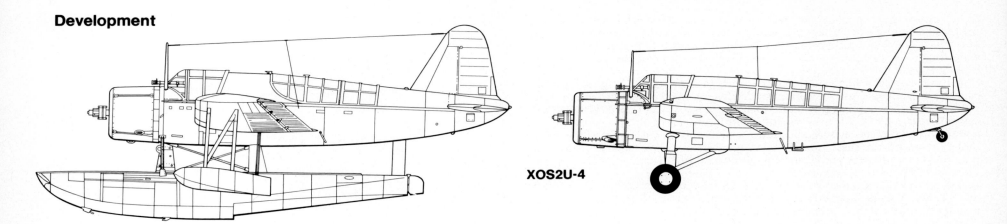

XOS2U-1

XOS2U-4

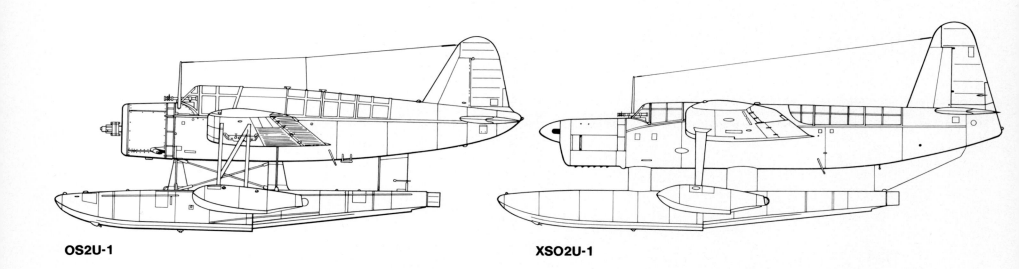

OS2U-1

XSO2U-1

OS2U-1

On 22 May 1939, BuAer ordered fifty-four OS2U-1s from Vought under contract number 66259 and assigned them BuNos 1681 thru 1734. These aircraft were delivered over a seven month period from May to December of 1940. The first aircraft were delivered to Observation Squadron Four (VSO-4) aboard the battleship USS COLORADO (BB-45).

Production OS2U-1s differed from the prototype in a number of ways. The aircraft was powered by a 450 hp Pratt & Whitney R-985-48 in place of the R-985-4 used on the XOS2U-1. The -48 engine used a longer exhaust stack than the -4. This stack protruded from the rear of the cowling. Dimensions and performance remained the same as the prototype; however, the OS2U-1 has a squared off leading edge to the observer's greenhouse canopy and had a Directional Finder (DF) loop installed in the rear cockpit. The centerline float mounting was changed with the rear support pylon being made somewhat wider and the floats were changed from the Vought designed units to ones built by the Edo Corporation which had a more rounded bow and different water rudder shape.

The OS2U-1 was armed with a fixed Browning .30 caliber machine gun with 500 rounds in the nose and a flexible .30 caliber machine gun in the rear cockpit with a total of 600 rounds of ammunition. The gunner also doubled as the radio operator and observer. The OS2U-1 was fitted with underwing racks (one per wing) that could accommodate either a 100 pound bomb or a 325 pound depth charge.

The production OS2U-1 had an empty weight of 4,643 pounds and its performance remained the same as the XOS2U-1 with a top speed of 175 mph at 5,500 feet. Service ceiling was 19,000 feet and range was 982 nautical miles in the float configuration. In the wheel configuration speed increased by some 2 mph.

Of the fifty-four OS2U-1s built, forty-nine were on Edo built floats and assigned to observation and scout duties aboard battleships and cruisers, while five were on wheel landing gear and assigned to land bases. The OS2U-1 was the first monoplane to serve aboard battleships, being assigned to the following Battleship Observation Squadrons (VO):

VO-1 with nine aircraft (Red tails), deployed detachments aboard USS ARIZONA (BB-39), USS NEVADA (BB-36) and USS PENNSYLVANIA (BB-38); VO-2 with nine aircraft (White tails), deployed detachments aboard USS TENNESSEE (BB-43), USS OKLAHOMA (BB-37) and USS CALIFORNIA (BB-44); VO-3 with nine aircraft (Blue tails), deployed detachments aboard USS IDAHO (BB-42), USS MISSISSIPPI (BB-41) and USS NEW MEXICO (BB-40); VO-4 with nine aircraft (Black tails), deployed detachments aboard USS WEST VIRGINIA (BB-48), USS COLORADO (BB-45) and USS MARYLAND (BB-46); and VO-5 with nine aircraft (Yellow tails), deployed detachments aboard USS TEXAS (BB-35), USS NEW YORK (BB-34) and USS ARKANSAS (BB-33). The PENNSYLVANIA served as the Fleet Flagship and the CALIFORNIA acted as the Flagship of the Battle Forces.

The first production OS2U-1 (BuNo 1681) was bailed (loaned) to Northrop Corporation of Hawthorne, California during 1941 for the installation of a set of experimental "Zap Flap" wings. Mr. Ed Zap, a close associate of Jack Northrop, had designed the full span flap wing with the intention of installing it on the production P-61 Black Widow. The wings were built and installed under contract number 75017 dated 25 September 1940. The new wings increased the OS2U-1's wing span to thirty-six feet six inches and decreased wing area to 218 square feet. The wings were of an all aluminum, alloy construction and there were no provisions for external stores. The "Zap" flaps were manually controlled and were set from the front cockpit only. The prototype XP-61 was fitted with this wing, but production tolerances precluded its use on the production P-61 model. A similar set of wings were installed on two OS2U-2s (BuNo 2189 and 3075) and these aircraft were redesignated as XOS2U-4. In the event, these wings were not fitted to production OS2Us.

On 1 October 1941, the OS2U was given the name Kingfisher by order of the Secretary of the Navy. The name Kingfisher would also be used by the Royal Navy Fleet Air Arm as the name for the OS2U-3s received under Lend-Lease.

The first production OS2U-1 (BuNo 1681) undergoes sea trials off of Norfolk, Virginia during 1940. The OS2U-1 proved to be an excellent rough sea aircraft, capable of handling waves in the five to nine foot range. BuNo 1681 was later shipped to Northrop for installation of a set of "Zap Flap" wings. (Navy)

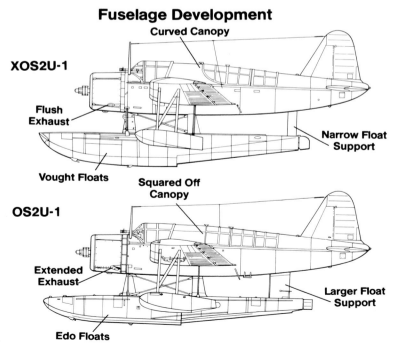

Fuselage Development

This OS2U-1 (BuNo 1692), in flight over Oahu, Hawaii, was assigned to Battleship Scouting Squadron One, aboard USS PENNSYLVANIA (BB 38). The aircraft is overall Aluminum Dope with Chrome Yellow wing uppersurfaces. The nose, tail, wing and fuselage stripes are Insignia Red. (National Archives)

The Battleship MISSISSIPPI (BB-41) was home to this OS2U-1 (BuNo 1714) of Observation Squadron Three (VO-3), enroute to NAS Alameda, California during late 1940. The aircraft was overall Aluminum Dope with Chrome Yellow wing uppersurfaces. The cowling was White, as was the fuselage band and upper wing chevrons. The tail assembly was painted True Blue. (Navy)

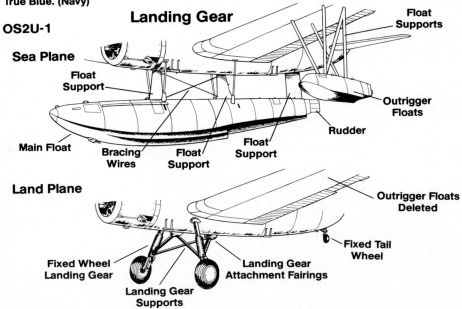

Landing Gear

OS2U-1

Sea Plane — Float Supports, Float Support, Outrigger Floats, Rudder, Main Float, Bracing Wires, Float Support

Land Plane — Outrigger Floats Deleted, Fixed Tail Wheel, Fixed Wheel Landing Gear, Landing Gear Attachment Fairings, Landing Gear Supports

An OS2U-1 makes an approach to a recovery sled during 1942. The Kingfisher was finished in Intermediate Blue Gray over Light Gray camouflage. The OS2U was a very stable sea plane and could handle rough seas. (LTV)

The Navy Aero Medical Research Laboratory used this OS2U-1 (BuNo 1693) for research on the G forces felt by pilots during dive pullouts. The aircraft was equipped with a spin recovery parachute housed under the lower rear fuselage. The aircraft was lost on 5 November 1942, when the pilot failed to pull out of a test dive. (National Archives)

The Admiral of the Fleet flew an OS2U-1 (BuNo 1691) during 1940-41 with a Dark Blue fuselage, Silver wings, horizontal and vertical stabilizer and Chrome Yellow wing upper-surfaces. When the aircraft was float equipped, the float was painted Aluminum Dope. (National Archives)

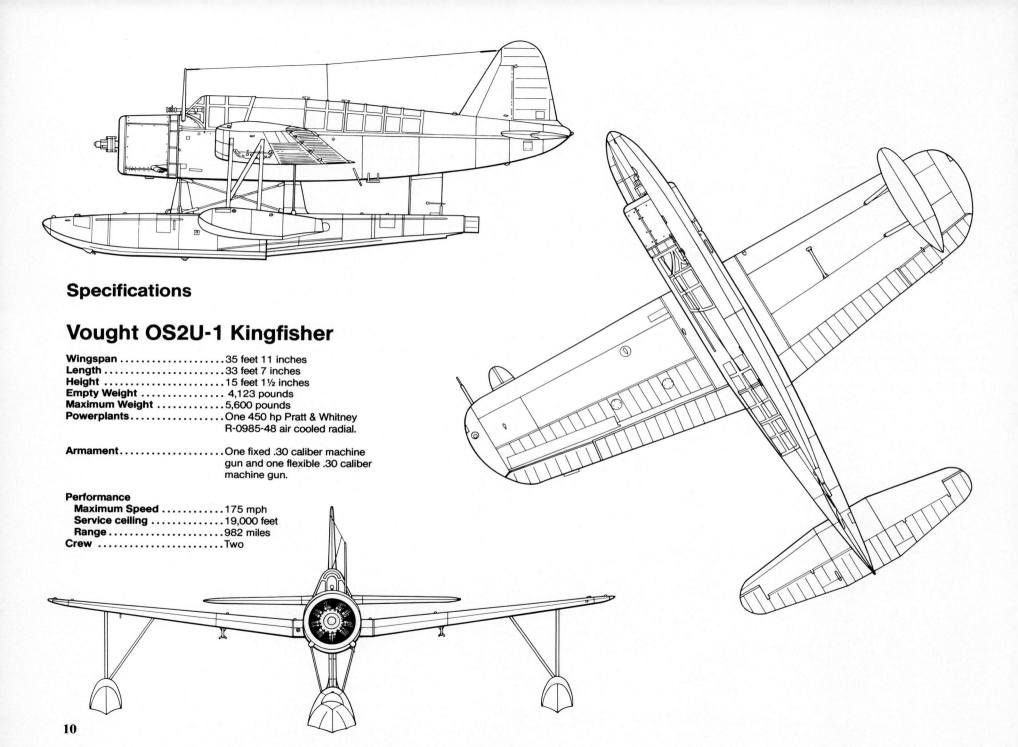

Specifications

Vought OS2U-1 Kingfisher

Wingspan	35 feet 11 inches
Length	33 feet 7 inches
Height	15 feet 1½ inches
Empty Weight	4,123 pounds
Maximum Weight	5,600 pounds
Powerplants	One 450 hp Pratt & Whitney R-0985-48 air cooled radial.
Armament	One fixed .30 caliber machine gun and one flexible .30 caliber machine gun.

Performance
- Maximum Speed .... 175 mph
- Service ceiling .... 19,000 feet
- Range .... 982 miles

Crew .... Two

The pilot's instrument panel was fitted with basic flying and engine instruments. The .30 caliber machine gun breech and ammunition can was in the pilot's compartment with the rudder pedals being placed on either side of the ammunition can. The electrical panel was located just behind the machine gun. (National Archives)

The observer/radio operator's compartment was armed with a Colt-Browning M2 .30 caliber machine gun. The gun was attached to a ring mount that could be elevated 90 degrees. The gunner was provided with 600 rounds of ammunition. (National Archives)

Observer's .30 Caliber Machine Gun

Gun Stowed

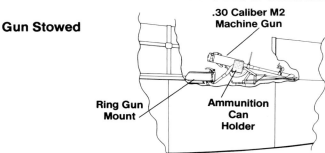

Gun Deployed

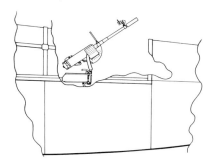

Three Kingfishers from Scouting Squadron One, First Naval District (VS-1D1) Squantum, Massachusetts patrol over the Atlantic Ocean during early 1942. Aircraft are camouflaged Intermediate Blue-Gray over Light Gray and are armed with 100 pound bombs. (LTV)

OS2U-1 (BuNo 1681) was fitted with a set of wings designed by Ed Zap, produced by Northrop and installed by Hardman Aircraft Products. These were of all metal construction and featured full span flaps with wing spoiler roll control. The wings were very similar to the wings installed on the later XOS2U-4, but could not be folded. The wing design was intended to be used on the Northrop XP-61 Black Widow. (LTV)

The auxiliary aileron or wing spoiler on the standard Kingfisher wing, provided lateral control when the ailerons were drooped with the flaps. With the flaps and ailerons in the fully drooped position, the OS2U could land at 55 mph. (National Archives)

The first production OS2U-1 (BuNo 1681) was used as a test bed to test the Northrop "Zap Flap" wing for possible use on Kingfishers. The wings were installed during 1941 and the aircraft underwent tests until 31 August 1943 when the aircraft was retired. (Navy)

OS2U-2

On 4 December 1939, even before Vought had finished production of the OS2U-1, the Navy issued a production contract for 158 slightly improved variants under the designation OS2U-2, to be produced in two batches (BuNos 2189-2288 and 3073-3130).

The OS2U-2 was externally identical to the earlier OS2U-1 differing in the installation of self-sealing fuel tanks, 140 pounds of armor and additional fuel tanks mounted in the inboard wing panels. Fuel capacity with the new tanks rose to 240 gallons, an increase of 96 gallons over the earlier OS2U-1. The fuel system was protected by a CO_2 gas purge system to help prevent any build up of explosive vapors. These changes brought the maximum weight of the OS2U-2 up to 6,108 pounds, causing a loss of some 7 mph in top speed, 170 mph for the OS2U-2, while the earlier OS2U-1 had a speed of 177mph.

The contract called for a total of 158 aircraft, with forty-five to be delivered as seaplanes and the remainder as land aircraft. Additionally, Edo supplied an additional 70 sets of floats allowing a portion of the land based aircraft to be easily converted to float configuration. All aircraft were to be powered by the P & W R-0985-50 engine.

The majority of the aircraft produced were delivered to either NAS Pensacola (46) or NAS Jacksonville (53), both air stations being located in Florida. These aircraft were used to form Inshore Patrol Squadrons. Eventually the Navy would form a total of thirty Naval District Inshore Patrol Squadrons, designating them as Scouting Squadrons (VS). VS squadrons could be either land or water based and operated from such bases as Squantum, Massachusetts (VS-1D1) to Coco Solo, Panama (VS-2D15) as well as bases in the Pacific. The following is a listing of these squadrons along with their new designations (which became effective on 1 March 1943):

This was the first production OS2U-2 (BuNo 2189) and was the first of 158 aircraft built by Vought during 1941. BuNo 2189 would later be fitted with a narrow chord wing and be used as a test bed. The OS2U-2 was essentially the same as the OS2U-1, with the major change being the addition of self-sealing fuel tanks, armored seats and the use of -50 Pratt and Whitney engine. (LTV)

VS-1D1 Squantum, MASS. (VS-31)	VS-2D1 Quonset, RI. (VS-32)
VS-3D1 Quonset, RI. (VS-33)	VS-1D3 New York, N.Y. (VS34)
VS-1D4 Cape May, N.J. (VS-35)	VS-5D4 Cape May, N.J. (VS-36)
VS-1D5 Norfolk, VA. (VS-37)	VS-2D5 Norfolk, VA. (VS-38)
VS-1D7 Banana River, (VS-39)	VS-2D7 Key West, FL. (VS-40)
VS-3D7 Key West, FL. (VS-62)	VS-1D10 San Juan, PR. (VS-63)
VS-2D10 Guantanamo, Cuba (VS-43)	VS-3D10 San Juan, PR. (VS-44)
VS-4D10 Trinadad (VS-45)	VS-1D11 San Pedro, CA. (VS-46)
VS-1D12 Alameda, CA. (VS-47)	VS-1D13 Terminal Island, CA. (VS-49)
VS-2D13 Seattle, WA. (VS-50)	VS-2D14 San Diego, CA. (VS-51)
VS-2D14 Quonset Point, RI. (VS-52)	VS-3D14 Pearl Harbor, HI. (VS-53)
VS-4D14 Alameda, CA. (VS-54)	VS-5D14 Alameda, CA. (VS-55)
VS-6D14 Alameda, CA. (VS-56)	VS-7D14 Alameda, CA. (VS-57)
VS-8D14 Alameda, CA. (VS-58)	VS-1D15 Coco Solo, Panama (VS-59)
VS-2D15 Coco Solo, Panama (VS-60)	

Initially, aircraft assigned to the inshore patrol units were painted in overall Aluminum dope with Yellow wing uppersurfaces and various colored cowlings and fuselage bands. Later these aircraft had a large national insignia applied to the nose to clearly identify the aircraft while conducting Neutrality Patrols.

Two battleships, USS NORTH CAROLINA (BB-55) and USS WASHINGTON (BB-56) did not carry numbered observation squadrons on board. NORTH CAROLINA received three OS2U-2s (BuNos 2288, 3073 and 3074) at the Brooklyn Navy Yard as ship's aircraft. These overall Light Gray aircraft all carried the legend NORTH CAROLINA on the fuselage sides in twelve inch White letters.

Besides being an observation and scout aircraft, the OS2U-2 was also used in the training role at NAS Jacksonville, NAS Pensacola and NAS Corpus Christi. These aircraft were used as intermediate and advanced trainers.

This OS2U-2 (BuNo 2190) served with Inshore Patrol Squadron 5 in the fourth Naval District (VS-4D1), located at Cape May, New Jersey. The Inshore Patrol Squadrons were designated as scouting (VS) units and formed to patrol the coasts in the anti-submarine/surface raider role. The aircraft was overall Aluminum Dope with Chrome Yellow upper wing surfaces. The cowling and fuselage stripe was Red. (LTV)

An OS2U-2 (BuNo 2228) taxies out into Pensacola Bay for a training flight. Aircraft number 30 was used in the intermediate and advanced sea plane training program. The aircraft was in the standard pre-war scheme of overall Aluminum with Chrome Yellow wing upper surfaces and Black markings. (National Archives)

With the advent of the Neutrality Patrol, Naval aircraft were marked with a large U. S. national insignia on the cowling as an identification aid to friendly gunners. This OS2U-2 (BuNo 2193) was stationed at Quonset Point, Rhode Island with VS-2D1. This unit would later become VS-32 on 1 March 1943. The nose and fuselage stripe was Red. (Naval History Center)

Beaching Gear

Vought Float

Edo Float

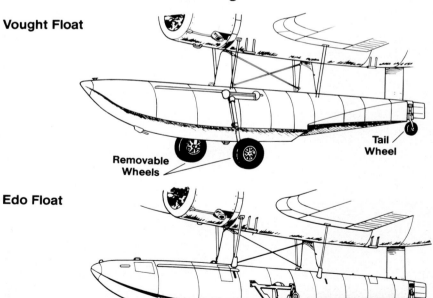

This OS2U-2 (BuNo 2288) was assigned to the battleship USS NORTH CAROLINA (BB-55) as part of her air group. The Kingfisher was overall Light Gray with White markings. The aircraft's centerline float and associated beaching gear were Vought-built rather than the standard Edo float. (LTV)

An OS2U-2 (BuNo 3102) during its acceptance trials on Long Island Sound. The aircraft is overall Light Gray which was the standard Navy scheme in use during 1941. Navy aircraft did not use this overall Gray scheme very long and in October of 1941, Intermediate Blue-Gray was added to the aircraft upper surfaces as a camouflage from above. (LTV)

One of the three OS2U-2s that were destined for the battleship NORTH CAROLINA undergoes final assembly at the Vought-Sikorsky plant. These OS2U-2s have the Vought streamlined float and beaching gear and were finished in overall Light Gray with White markings. (National Archives)

One of the aircraft's crew sits on the float of this capsized OS2U-2 which lost its starboard wing outrigger float. The aircraft was recovered and given a fresh water wash down. The instruments and engine were replaced and the aircraft was returned to service. (National Archives)

The seaplane tender USS TANGIER (AV-8, ex-SS Sea Arrow) off loads an OS2U-2 Kingfisher as a rescue launch stands by. The ship was an ex-C-3 cargo ship that was converted to a seaplane tender during 1941. The purpose of the tender was to provide a mobile maintenance facility for the aircraft at sea. (National Archives)

The beaching crew is waving a Green flag to indicate to the pilot that this is the proper beaching ramp. Recovery during the summer months would be a pleasant experience, giving the crew a chance to cool off. The wheels in the foreground are the beaching gear. (National Archives)

An OS2U-2 is launched from the seaplane tender USS TANGIER (AV-8) in the Pacific during early 1942. The launching crewman in front of the port float is indicating to the crane operator to lift the OS2U-2 clear of the deck. The aircraft on the deck are resting on beaching gear, which were removed after the crane lifted the aircraft. (National Archives)

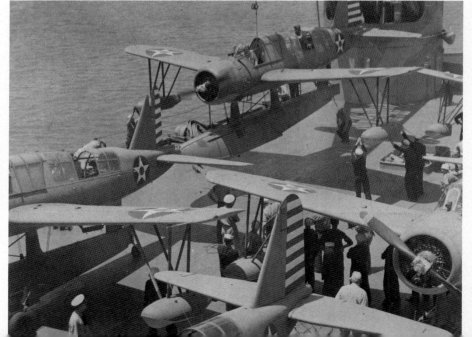

Recovery operations in Alaska during the winter months was a harsh experience. This OS2U-2 was being recovered at Womens Bay, Kodiak, Alaska during 1943. The aircraft had just returned from a patrol mission out over the Aleutian Islands. (Edo Corp)

The beaching gang attaches the beaching gear to this OS2U-2 at NAS Corpus Christi, Texas during 1943. The aircraft is equipped with practice bomb dispensers under each wing on the standard bomb rack. (National Archives)

The Commander of Patrol Wings, Atlantic Fleet, RADM A.D. Bernhard flew this OS2U-2 Kingfisher from NAS Norfolk, Virginia during early 1942. The Red and White tail stripes were deleted during May of 1942. (Navy)

The aviation machinist mate checks out the engine of an OS2U-2 on the ramp at Naval Air Station Pensacola during 1943. The OS2U-2 was used as an intermediate and advanced seaplane trainer in the Naval Aviation Cadet training program. Training with the Kingfisher also took place at Jacksonville, Florida and Corpus Christi, Texas. (National Archives)

With the engine running, the crew of this Kingfisher await the arrival of a parachute before beginning their training mission. The boarding ramp was provided so the crew could get into and out of the aircraft without wading. The forward cockpit has had the canopy removed and the aircraft is armed, unusual for a trainer. The aircraft carries no fuselage national insignia, only the numbers 86 in Black. (National Archives)

Ground crews have folded the antenna over as they prepare to lift a well worn OS2U-2 from the water at NAS Pensacola during 1943. The aircraft will be lifted and the beaching gear installed while the aircraft is suspended. Salt water exposure has caused severe weathering on all painted surfaces. The aircraft number 65 is White and the stripe on the float is Red. (National Archives)

OS2U-3

The OS2U-3 was the most widely produced variant of the Kingfisher with over 970 being built under contract C76493. The first OS2U-3 was accepted by the Navy at Anacostia on 23 May 1941. The OS2U-3 production run was given the BuNos 5284-5941, 5973-5989 and 09393-09692.

The aircraft was powered by a 450 hp Pratt & Whitney R-985-AN-2 radial engine, the AN designation signifying Army/Navy. The AN designation indicated that the engine was common to both services and parts could be used interchangeably. This practice cut down on the confusion of different engine designations (for the same power plant) for each service and eased parts ordering for both.

The OS2U-3 was externally identical to the earlier OS2U-1 and -2. With the addition of extra crew armor, the empty weight rose to 4,560 pounds. Performance was similar to earlier variants with a maximum speed at 5,000 feet of 171 mph, a cruising speed of 152 mph and a landing speed of 55 mph. The aircraft had a rate of climb of 800 feet per minute and a range of 908 miles. This range gave the OS2U-3 an endurance of some six hours.

The OS2U-3 was equipped with a bracket on the starboard side of the forward nose adjacent to the windshield for mounting a Fairchild gun camera. The camera itself was bulky and cumbersome and was never used in combat, being installed only for training and gunnery practice. As with previous variants the OS2U-3 also used the MK III telescopic gun sight.

The OS2U could be equipped with a Fairchild 35MM gun camera on a mounting alongside the windscreen. The camera was activated when the .30 caliber machine gun was fired. In actual practice the camera was used only for training purposes and not in combat. (National Archives)

The OS2U-3 was the most numerous variant of the Kingfisher series with some 1,006 aircraft being built by Vought-Sikorsky. This aircraft (BuNo 5284) was the first production OS2U-3 and was rolled out in the then standard overall Light Gray scheme with White markings. (LTV)

The utility hangar at NAS Norfolk housed at least twelve Kingfishers and three Grumman Ducks during early 1942. The OS2U-3 Kingfishers are serving with VS-1D5, while the J2F Ducks are assigned to Utility Squadron Four (VJ-4). (National Archives)

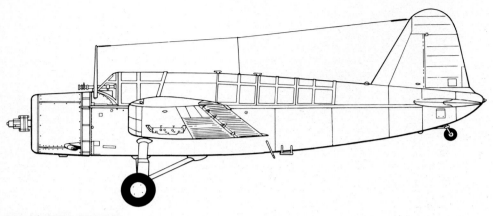

Specifications

Vought OS2U-3 Kingfisher

Wingspan	35 feet 11 inches
Length	33 feet 7 inches
Height	15 feet 1 1½ inches
Empty Weight	4,310 pounds
Maximum Weight	6,108 pounds
Powerplant	One 450 hp Pratt & Whitney R-985-AN-2 air cooled radial.
Armament	One fixed .30 caliber machine gun and one flexible .30 caliber machine gun.

Performance
   Maximum Speed .... 171 mph
   Service ceiling .... 15,500 feet
   Range .... 908 miles
Crew .... Two

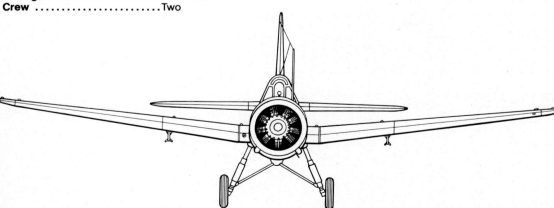

The crew chief of this OS2U-3 warms up the engine while the pilot signs off the maintenance log. The aircraft was assigned to VS-1D5 based at NAS Norfolk, Virginia. The Kingfisher is armed with two 325 pound depth charges for anti-submarine patrol work. (National Archives)

Aircraft ordnancemen wheel out two 100 pound bombs to an OS2U-3 (aircraft number 18) on a torpedo cart. Although the crewmen are plainly getting ready to load this OS2U-3, the wartime photo censors have painted out the underwing bomb racks. (Archive United Technologies)

Pilots and observers walk to their waiting Kingfishers at NAS Norfolk, Virginia during early 1942. OS2U-3s were credited with sinking two German U-Boats, one off the coast of North Carolina during 1942, and the other off the coast of Cuba during 1943. (National Archives)

Bomb Rack

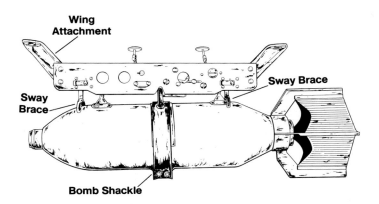

This OS2U-3 Kingfisher (BuNo 9475), aircraft number 18, begins its takeoff run armed with two 100 pound bombs. Operating from NAS Norfolk in 1942, the six hour cruising endurance of the OS2U-3 gave it a wide scouting area. This OS2U-3 was assigned to VS-1D5 which patrolled off the Atlantic Coast looking for German submarines. (Archives United Technologies)

An OS2U-3 (BuNo 5865) armed with 100 pound bombs flies over the Atlantic on an anti-submarine patrol. The aircraft was camouflaged in Intermediate Blue Gray uppersurfaces over Light Gray undersurfaces with Black numbers. The recovery net hook and catapult fitting are clearly visible on the float keel. (National Archives)

The plane captain of this OS2U-3, aircraft number 71, waits while the aircraft is run through a fresh water wash at NAS Corpus Christi, Texas. The aircraft was overall Aluminum Dope with Chrome Yellow upper wing surfaces. The fresh water sprayed helped prevent corrosion from the salt water. (National Archives)

The observer has stowed the rear two sections of the canopy and brought his .30 caliber machine gun out into position for firing. He had 600 rounds of ammunition for the flexible gun and the pilot had 500 rounds for the fixed forward firing .30 caliber machine gun. (National Archives)

The student pilot of this OS2U-3 approaches the target sock from behind and below for the best possible angle on the target. These maneuvers were meant to sharpen the gunnery skills of the cadets during their training at NAS Pensacola. The aircraft is finished in overall Aluminum Dope with Chrome Yellow upper wing surfaces. (National Archives)

Four OS2U-3s fly in a right echelon formation over Padre Island, Corpus Christi, Texas during late 1942. The instructor pilot is leading the flight from aircraft number 100. The aircraft were painted in Intermediate Blue Gray uppersurfaces over Light Gray undersurfaces with Chrome Yellow upper wing surfaces. (Archive United Technologies)

An OS2U-3 is brought aboard a sea plane tender at Cooks Inlet, Anchorage, Alaska, for repairs to the port wing. The damage was received during a spotting mission for the USS SALT LAKE CITY (CA-25) off Attu Island, during March of 1943. The Japanese came very close to sinking the heavy cruiser, but luckily the Japanese commander withdrew due to a lack of ammunition. (National Archives)

An OS2U-3 is hoisted aboard a battleship after a scouting mission during 1943 as the carrier YORKTOWN (CV-10) steams past in the background. Kingfishers served aboard ship until they were gradually replaced by the Curtiss SC-1 during late 1944. This OS2U-3 is armed with two 100 pound bombs on the wing racks. (National Archives)

Row upon row of OS2U-3s await their turn to fly at NAS Corpus Christi, Texas during 1942. Two color schemes are evident: the overall Aluminum Dope and the Intermediate Blue Gray over Light Gray. Trainer aircraft (in either paint scheme) still employed Chrome Yellow upper wing surfaces. (National Archives)

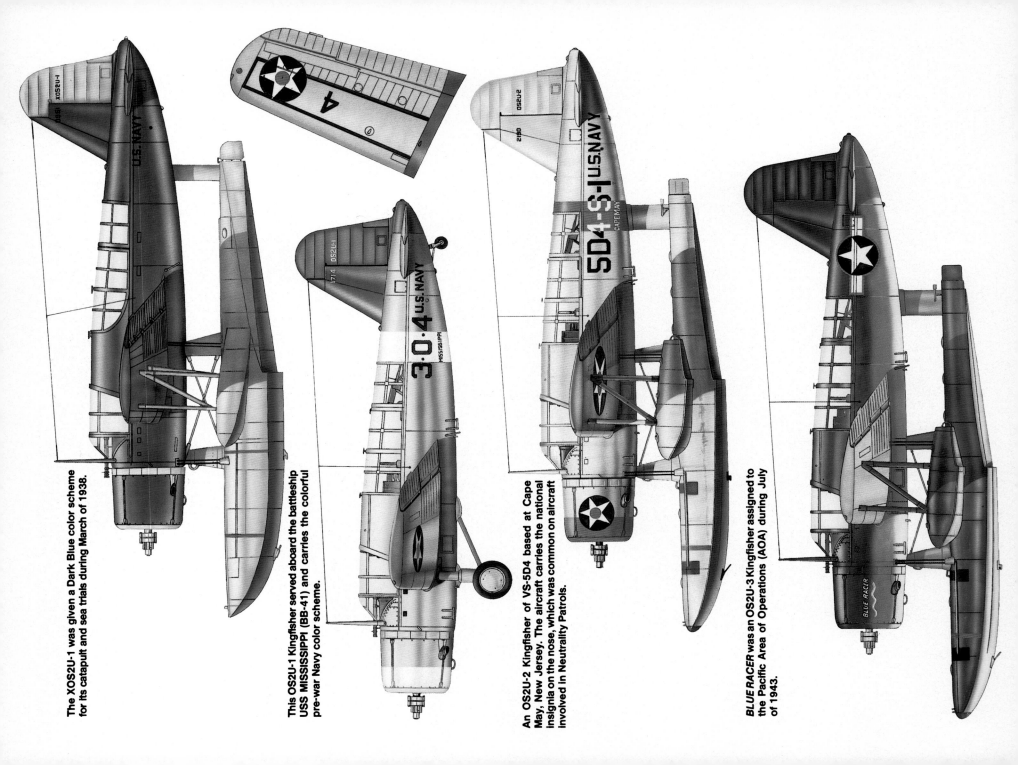

The XOS2U-1 was given a Dark Blue color scheme for its catapult and sea trials during March of 1938.

This OS2U-1 Kingfisher served aboard the battleship USS MISSISSIPPI (BB-41) and carries the colorful pre-war Navy color scheme.

An OS2U-2 Kingfisher of VS-5D4 based at Cape May, New Jersey. The aircraft carries the national insignia on the nose, which was common on aircraft involved in Neutrality Patrols.

BLUE RACER was an OS2U-3 Kingfisher assigned to the Pacific Area of Operations (AOA) during July of 1943.

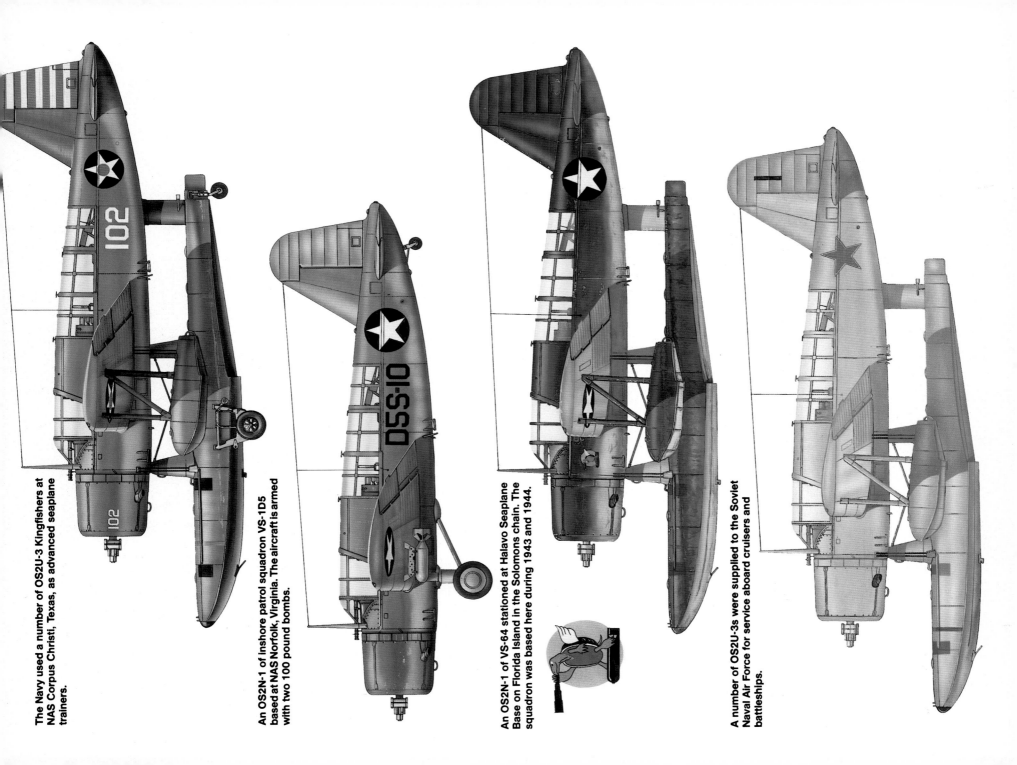

The Navy used a number of OS2U-3 Kingfishers at NAS Corpus Christi, Texas, as advanced seaplane trainers.

An OS2N-1 of inshore patrol squadron VS-1D5 based at NAS Norfolk, Virginia. The aircraft is armed with two 100 pound bombs.

An OS2N-1 of VS-64 stationed at Halavo Seaplane Base on Florida Island in the Solomons chain. The squadron was based here during 1943 and 1944.

A number of OS2U-3s were supplied to the Soviet Naval Air Force for service aboard cruisers and battleships.

Beach crews from VS-64 prepare an OS2U-3 for flight from Halavo Seaplane Base, Florida Island in the Solomons chain during July of 1943. The salt water and high humidity conditions play havoc with the aircraft and they had to be constantly cleaned to prevent corrosion. (National Archives)

This OS2U-3 of VS-64 was undergoing a major overhaul at Halavo Seaplane Base during December of 1943. Particular attention was being given to the Pratt and Whitney 450 hp R-985-AN2 engine. The conditions at this advance seaplane base are less than satisfactory for both aircraft and crews. (National Archives)

Two Navy pilots discuss an upcoming training mission under the wing of an OS2U-3 at NAS Corpus Christi, Texas. The aircraft would be towed to the beach and a crane would lift the aircraft into the water once the beaching gear was removed. (National Archives)

An OS2U-3 flies over Angaur, Palau Island, on 15 September 1944, while elements of the 1st Marine Regiment move ashore to capture the island. The Kingfisher was spotting gunfire for the cruiser USS PORTLAND (CA-33) which was firing on entrenched Japanese gun positions on the island. (LTV)

Aviation Machinists Mates prepare a sealer to be painted on the floats of the Kingfisher in the background. The Kingfisher in the foreground has already been treated with the sealer. The aircraft catapults were mounted on turntables that could rotate 360 degrees. (National Archives)

As Sunday church services are being held on the fan tail, an OS2U-3 Kingfisher sits on the ship's catapult waiting for its next scouting mission. The supports that go from the catapult to the wing float struts are used to stabilize the aircraft in the event of rough seas or strong winds. The tank on the fan tail contains aviation gasoline. (National Archives)

An OS2U-3 is launched off of the starboard catapult of a battleship. The aircraft is finished in the three-tone camouflage scheme introduced during early 1943. The catapult was gun powder driven and propelled the Kingfisher from 0 to 70 mph, a speed sufficient for safe flight. (Edo Corp)

The catapult officer prepares to give the launch signal as the Kingfisher pilot brings the engine up to full throttle. The OS2U-3 is equipped with an additional antenna in front of the rear cockpit for the short wave band radio and is carrying practice bomb dispensers. (National Archives)

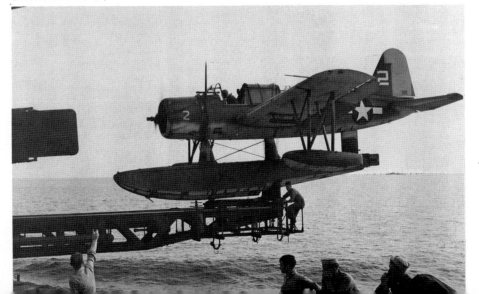

Catapult

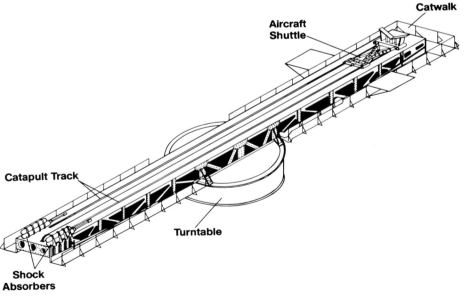

In an attempt to increase the lift of his Kingfisher, LT T.W. Kerker of VS-68 designed and installed these lengthened wing tips. The wing material was taken from non-flying aircraft on Ugi Island during November of 1943. There is no record of flight tests or authorization to modify the Kingfisher. (National Archives)

A chilly night recovery of an OS2U-3 of Scouting Observation Service Unit Two (SOSU-2). The scouting utility squadrons were formed to train pilots for combat after they graduated from flight school. This Kingfisher was finished in the three tone camouflage and carried the individual aircraft code 2V7 on the cowling in White. (Edo Corp)

A crew chief warms up the engine on an OS2U-3 Kingfisher serving with Scouting Observation Service Unit Two (SOSU-2). SOSU Units performed rear echelon maintenance on the Kingfishers as well as training of newly graduated cadet pilots. The aircraft was finished in the three tone camouflage paint scheme authorized during early 1943. (Edo Corp)

A Kingfisher coded 2-SU-3 of SOSU-2 lands after a training flight. The aircraft is camouflaged in the three tone scheme and is configured with practice bomb dispensers. The brackets for the Fairchild serial gun camera are installed just below the pilot's windshield and the aircraft is equipped with flame dampers on the exhaust stubs. (Edo Corp)

The pilot of this Kingfisher is maneuvering into position to catch the recovery sled as the recovery crewman displays a Green flag. This rather rough water recovery took place on 29 April 1943, aboard the USS SOUTH DAKOTA during operations in the Pacific. (Naval History Center)

Aircraft number 2 from an unidentified scouting squadron approaches the recovery sled. As soon as the hook on the float engages the sled, the lifting cable will be attached and the OS2U-3 will be brought aboard. The Kingfisher was camouflaged in Gloss Sea Blue on the upper surfaces over Insignia White lower surfaces. The emblem on the tail is Donald Duck with a telescope. (National Archives)

An OS2U-3 approaches the recovery sled for a starboard recovery during 1944. The ship has made a sweeping turn to starboard to create an area of calm water for the Kingfisher to land on. This recovery sled consists of canvas and a six foot square of cargo netting. Recovery operations were usually accomplished as the ship steamed at normal operating speeds. (National Archives)

Rescue Work

Although the Kingfisher was designed as a scout and observation aircraft, it is best remembered as a rescue aircraft, carrying downed aircrews in the gunner's compartment or clinging to the wing.

One of the most famous Kingfisher rescue missions was performed by LTJG F.E. Woodward and AR1 L.H.Boutte on 11 November 1942. The Kingfisher crew found and rescued four men that day, including CAPT Eddie Rickenbacker, the top scoring American Ace of the First World War. CAPT Rickenbacker's aircraft had been forced down in late October 1942 and the crew was adrift for some three weeks before being spotted by LT Woodward. Unable to put all the men into the aircraft, named the *Bug*, LT Woodward lashed Rickenbacker and another man to the wing and taxied some forty miles to land. The *Bug* was a charmed aircraft (BuNo 2201) having survived the Pearl Harbor raid aboard USS PENNSYLVANIA with only minor damage.

Another rescue took place off Truk Island on 30 April 1944 by two OS2Us off the USS NORTH CAROLINA (BB-55). The Kingfishers had been launched in an effort to locate a downed fighter pilot off USS ENTERPRISE, LT Bob Kanze. He was spotted drifting in his life raft in heavy seas and LT N. Dowdle landed his Kingfisher and taxied toward the downed pilot. The heavy seas, however, caused the Kingfisher to capsize and now there were three men in the water. The other orbiting Kingfisher, flown by LT John Burns, landed and taxied to the three men who were clinging to LT Kanze's raft.

The Kingfisher also operated as a search and rescue aircraft saving downed American flyers. LTJG G. M. Blair's Hellcat from VF-9 was downed in Truk lagoon on 23 March 1944. He was picked up and given a ride back to the ship in the observers position of this Kingfisher. Aviation Chief Radioman R.F. Hickman stands on the wing to catch the hoisting cable. (Naval History Center)

Although specifically designed as a scout observation aircraft, the OS2U was pressed into service as a search and rescue aircraft. Truk Lagoon was the scene of an unparalleled rescue of 22 airmen on 30 April 1944 by two Kingfishers and the submarine USS TANG (SS 306). LT John Burn's Kingfisher has eight airmen on the wings and fuselage. (National Archives)

After the three men were aboard the Kingfisher, LT Burns taxied the aircraft to the submarine USS TANG (SS-306). To prevent LT Dowdle's capsized OS2U from falling into enemy hands, the gun crew of TANG sank the aircraft. Later that same day TANG would rescue LT Burns after his Kingfisher was damaged in another rescue mission. On that mission, Burns had rescued seven other downed airmen. Once more TANG's gunners sank the damaged OS2U to keep it out of Japaneses hands. In all, TANG rescued some twenty-two men that day.

In another incident, a Kingfisher also scored a kill on a Japanese fighter. On 16 February 1945, LTJG D.W. Gandy was spotting gunfire for USS PENSACOLA off Iwo Jima when he was attacked by a Japanese fighter. Luckily the Japanese pilot missed in his attack and as he was turning to make another pass, LT Gandy maneuvered behind him and opened fire with his forward machine gun. He scored hits on the wing and engine area. He continued to fire until the Zero burst into flames and crashed onto the island.

Sailors from a torpedoed British freighter lay out on the wings of an OS2U-3 Kingfisher during 1944. Kingfisher pilots performed a number of daring rescues, picking up downed airmen and sailors while under fire. (LTV)

Foreign Service

England

The OS2U-3 was the only Kingfisher variant to be exported. The Royal Navy acquired 100 aircraft under Lend-Lease during 1942 assigning them the name Kingfisher I. The aircraft, BuNos 5811-5840 and 09513-09582, were shipped by sea to Scotland where they were reassembled and assigned British serials (FN650-FN749).

Before the Kingfisher I was accepted for Royal Navy service, three aircraft were tested. Two land versions were extensively tested at the Aeroplane and Armament Experimental Establishment (FN656 and FN651) between May and October of 1942, while a float version (FN678) was tested at the Marine Aircraft Experimental Establishment in Scotland during May of 1942. This aircraft was used to conduct sea worthiness trials. The results of the tests were successful and the Kingfisher I was accepted into Fleet Air Arm service.

The first Kingfisher squadron was No 703 Squadron formed at Lee-on-Solent during May of 1942. The squadron provided aircraft detachments (flights) to the Armed Merchant Cruisers which operated in the South Atlantic and Indian Oceans and to the cruisers HMS ENTERPRISE and HMS EMERALD. Two squadrons were established to act as training units, No 764 at Lawrenny Ferry and No 765 at Sandbanks (Poole) Harbor, Dorset. These squadrons trained Royal Navy pilots on the Kingfisher.

Two squadrons were formed as Fleet Requirements Units, towing targets and performing other duties such as radar calibration. Both squadrons were based in South Africa. One unit, No 726 Squadron, at Durban and the other, No 789 Squadron, at Wingfield. Another training unit, No 740 Squadron, acted as a land based training squadron operating out of Arbroath, England.

By early 1944, the majority of Kingfisher Is had been phased out of Fleet Air Arm service. Twenty were returned to the U.S. Navy and it is believed that several were given to the Soviet Union for use with the Soviet Naval Air Force.

Australia

The Royal Australian Air Force (RAAF) received eighteen aircraft that had been relinquished from an order for twenty-four aircraft that were enroute to the Dutch East Indies when the islands fell to the Japanese on 9 March 1942. The aircraft were given the RAAF serials A48-1 through A48-18. The first RAAF squadron was No 105 Squadron which was formed on 10 May 1943 at RAAF Station Rathmines. While based at Rathmines the squadron lost three Kingfishers in accidents. The squadron moved to Saint Georges Basin, New South Wales during January of 1944 and remained there until it was disbanded on 31 October 1945. The surviving aircraft were then placed in storage.

During their service in Australia, several Kingfishers of No 107 Squadron were used in a series of camouflage experiments. Initially, when the Kingfishers first arrived in Australia, they were painted in overall Light Gray with Netherlands East Indies markings. This camouflage was deemed unacceptable for conditions in Australia and a new scheme was applied. Several different camouflage colors and patterns were experimented with before the aircraft were finally camouflaged in an upper surface pattern of Extra Dark Sea Gray and Dark Slate Gray with Sky Blue undersurfaces. The squadron code of No 107 Squadron (JE) and the individual aircraft letter were carried on the fuselage in Medium Sea Gray.

During November of 1947, one Kingfisher (A48-13) was transferred from the RAAF to the Australian Department of External Affairs for service with the Australian contingent in Antarctica. The aircraft was painted a high visibility scheme of overall Orange Yellow and embarked in the supply ship HMAS WYATT EARP.

Other Exports

Kingfishers were exported to Chile, which received fifteen aircraft (BuNos 5911-5925) under Lend-Lease (one is now on display at the air force museum). Other Latin American nations that received OS2U-3 Kingfishers included Uruguay (6), Argentina (9), Mexico (6) and the Dominican Republic (3). The three aircraft that were delivered to the Dominican Republic were reportedly later transferred to Cuba. These aircraft were reportedly used by rebels to bomb government troops during the Castro rebellion of 1958. At least one of these aircraft is currently on display in Cuba.

The Fleet Air Arm found the OS2U-3 vastly superior to the Curtiss SO3C Seagull and actually refused delivery of the last batch of Seagulls during 1943. This OS2U-3 (FN668, ex-BuNo 5830) was used aboard the AMC CILICIA for one year while operating in the South Atlantic searching for German and Japanese merchant raiders. (National Archives)

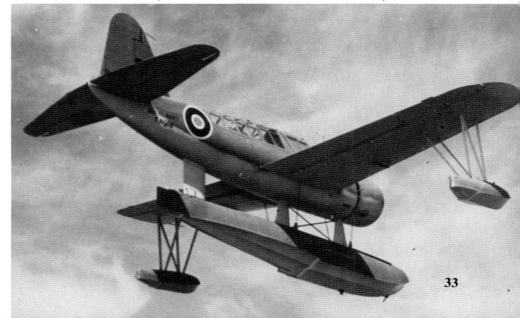

The British were supplied with 100 OS2U-3 Kingfishers under Lend-Lease to be used as trainers, scouts and observation aircraft. This OS2U-3 (BuNo 5817) finished in Intermediate Blue Gray over Light Gray would become a Kingfisher I, Fleet Air Arm serial FN656 and be used as a test aircraft at the Aeroplane and Armament Experimental Establishment during late 1942. (LTV)

A Kingfisher I (FN668, ex-BuNo 5830) flies over the English Channel during late 1942. The aircraft was finished in Intermediate Blue Gray over Light Gray, the same scheme as used by the U.S. Navy. The Royal Navy Fleet Air Arm used the Kingfisher for two years, both aboard ship and at various shore installations. (National Archives)

A Kingfisher I (OS2U-3, ex-BuNo 5833) of the Royal Navy Fleet Air Arm is lowered into the water off HMS PEGASUS. The ship operated as the R.N. training ship for catapult operations. Pilots and observers that trained on the PEGASUS went on to serve aboard Armed Merchant Cruisers (AMC) and ships of the line. The observer usually rode on top of the fuselage to help balance the aircraft. (LTV)

The HMS PEGASUS operated as the Royal Navy catapult training ship during the Second World War. This OS2U-3 Kingfisher I (FN672, ex-BuNo 5833) is being lowered to the sea in preparation for another training mission. There is a cartoon character of Donald Duck on the cowling. (LTV)

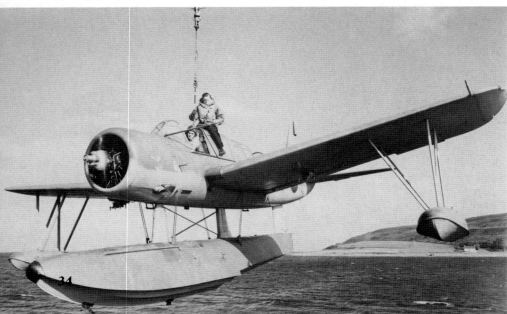

Gunsights

OS2U-3

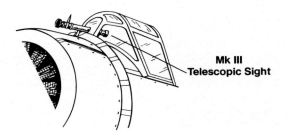

Mk III Telescopic Sight

Kingfisher I (OS2U-3)

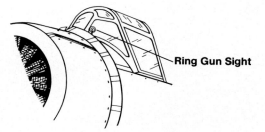

Ring Gun Sight

This OS2U-3 (A48-13) was one of a group intended for the Dutch East Indies and diverted to Rathmines, Australia when the Indies fell. The aircraft was assigned to No 107 Squadron and finished in overall Light Gray with all lettering in Black. (RAAF Museum via D.W. Gardner)

Chile received fifteen OS2U-3s under Lend-Lease during 1943. The aircraft (BuNos 5911-5925) were used by the Chilean Air Force (FAC) to patrol the long coast line of Chile and for anti-submarine duties. (Museo de Chile)

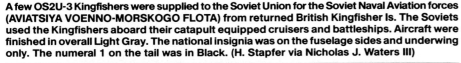

A few OS2U-3 Kingfishers were supplied to the Soviet Union for the Soviet Naval Aviation forces (AVIATSIYA VOENNO-MORSKOGO FLOTA) from returned British Kingfisher Is. The Soviets used the Kingfishers aboard their catapult equipped cruisers and battleships. Aircraft were finished in overall Light Gray. The national insignia was on the fuselage sides and underwing only. The numeral 1 on the tail was in Black. (H. Stapfer via Nicholas J. Waters III)

Uruguay received eight OS2U-3 Kingfishers under Lend-Lease during 1943. The OS2U-3 was finished in Intermediate Blue Gray over Light Gray with the lettering in White. A-752 (Armada-752) was preserved and placed on blocks for display outside the main naval base in Uruguay. (Nicholas J. Waters III)

XOS2U-4

Two OS2U-2s (BuNos 2189 and 3075), were selected by BuAir to be modified with a set of experimental "Zap Flap" wings.

The wings, designed by Ed Zap of the Northrop Corporation, were of an all aluminum alloy construction with full span flaps and upper wing spoilers for lateral control. The wings were very similar to the ones that had been installed earlier on an OS2U-1 (BuNo 1681) except that these wings could be folded in much the same manner as the wings on the Grumman TBF Avenger and Vought-Sikorsky XSO2U-1.

Two different manufacturers were chosen to install the wings. Vought-Sikorsky worked on BuNo 2189 and Douglas Aircraft received BuNo 3075. Vultee Aircraft was the contractor for the wing installation on the Douglas aircraft.

After the wings had been installed by each company, the aircraft were re-designated as XOS2U-4s and both were flown to NAS Moffett Field, California for extensive flight tests. These tests took place for over a two year period and considerable data was gathered on the performance of the wing.

The wing design, although tested on the XP-61, was never put into production during World War II due to the high production tolerances required by the design and the XOS2U-4 was cancelled. BuNo 3075 was stricken from the Navy inventory on 31 March, 1945, with BuNo 2189 being retired exactly one year later.

The horizontal stabilizers were also modified with a tapered chord and squared off tips.

Two OS2U-2 aircraft were fitted with a set of experimental "Zap Flap" wings and redesignated as XOS2U-4s. The wings featured full length flaps that increased wing area when extended by twenty-five percent. No changes were made to the fuselage with the exception of the addition of an anti-spin parachute in the tail cone. (Navy)

For flight tests, an anti-spin parachute was installed in the tail cone of the XOS2U-4. The chute was deployed when and if the pilot felt that the aircraft might go into a spin or when the aircraft had already entered into a spin. (National Archives)

Both XOS2U-4s were configured as land planes since neither had provisions for the installation of wing tip floats on the modified wing. The XOS2U-4 was powered by a 450 hp Pratt and Whitney R-985-50 engine. (Navy)

The XOSU-4 provided the US Navy and aircraft manufacturers with aeronautical information that in later years would prove to be invaluable. The "Zap Flap" wing and spoiler design was so far advanced from the conventional wings of the early 1940s that it proved impossible to mass produce the wings because of their close production requirements. (Navy)

The "Zap Flaps" enabled the XOS2U-4 to land at 55 mph and significantly reduced the take-off roll. Aileron control was supplied by spoilers on the upper wing. The all metal wing had a span of thirty-six feet six inches. (Navy)

Both XOS2U-4s (BuNo 2189 and 3075) were painted in the then standard Intermediate Blue-Gray over Light Gray color scheme. After various tests at NAS Norfolk and NAS Anacostia, both aircraft were flown to Moffett Field for extensive tests of the new wing. In the event, production tolerances prevented its use during World War II. (Navy)

OS2N-1

During 1942, production of the OS2U was transferred to the Naval Aircraft Factory (NAF), Philadelphia Navy Yard, Pennsylvania, to free the assembly lines at Vought-Sikorsky for production of the Chance Vought F4U Corsair. On 9 March 1942, the first production OS2U-3 designated the OS2N-1, was accepted by the Navy (the N designation reflected the new manufacturing site, since N was the Navy manufacturers code letter for the NAF). The NAF produced some 300 OS2N-1s (BuNos 01216 - 01515) in less than a year and there are indications that the NAF actually assembled more than the 300 aircraft under contract. Vought, after they finished production of the OS2U-3, turned over enough spare parts to the NAF to produce another thirty-one aircraft in excess of the original contract (BuNos 5942-5972).

An OS2N-1 sits on a production roll around dolly at the Naval Aircraft Factory, Philadelphia, Pennsylvania. The Pratt and Whitney R-985-AN-2 had already been installed; however, the weapons and radio equipment would be fitted by the Navy after the aircraft was delivered to its respective squadron. (National Archives)

The Naval Aircraft Factory, located at the Philadelphia Navy Yard, Pennsylvania produced 300 OS2U-3s under the designation OS2N-1. Although the aircraft were identical in every way to the OS2U-3, the designation was changed to reflect the new manufacturing site. This OS2N-1 (BuNo 01254) awaits delivery to a Navy unit and is finished in Intermediate Blue Gray over Light Gray. (National Archives)

An OS2N-1 parked and chocked on the NAF ramp prior to delivery to a front line squadron. The aircraft in the background appears to be the XSBN-1, the Naval Aircraft Factory variant of the Brewster SBA-1 scout bomber. (National Archives)

The OS2N-1, like the OS2U-3, could be fitted with either fixed wheeled landing gear or floats. This OS2N-1 had a mismatched painted cowling and a field installed whip antenna on the fuselage spine just in front of the observer greenhouse. There is evidence that 31 additional OS2N-1s were assembled from spare parts provided by Vought. (National Archives)

Crewmen from VS-1D3 based in New York practice deploying the onboard life raft from this OS2N-1 during 1942. The life raft was stowed directly behind the pilot and was reached through a hatch in the bulkhead behind the pilot's head. VS-1D3 later became Scouting Squadron Thirty Four (VS-34) during March of 1943. (National Archives)

An OS2N-1 waits on the flight line at the Naval Aircraft Factory (NAF), Philadelphia Navy Yard for delivery with the aircraft's documents secured in a plastic bag hanging from the wing. By law, the NAF could produce up to ten percent of the aircraft used by Naval Aviation units. (National Archives)

An OS2N-1 flies over the coast on a scouting mission. The aircraft is finished in Intermediate Blue Gray over Light Gray. Wing tip floats, as on all OS2U series, were interchangeable from right to left. The catapult fitting and recovery hook are visible on the float keel. (National Archives)

An OS2N-1 of VS-50 is refueled and made ready for another anti-submarine patrol from Kodiak, Alaska. The aircraft is armed with 325 pound depth charges and the wing tie downs are 55 gallon drums filled with water. The wind on the Alaskan coast, known as the Williwaw, could gust up to 100 mph without warning. The Kingfisher is fitted with an extra antenna that extends from the fuselage side to the horizontal stabilizer. (National Archives)

Crewmen from VS-64 carry out maintenance on an OS2N-1 Kingfisher at Halavo Seaplane Base, Florida Island, Solomons Islands chain. The OS2N-1 is undergoing its 60 hour check that included an oil change and cleaning/inspection of corrosion prone components. The Squadron artist was applying the VS-64 emblem on the fuselage which consisted of a turtle with a telescope. (National Archives)

An OS2N-1 of VS-1D5, armed with 325 pound depth charges, flies over a convoy in the Atlantic during 1942. VS-1D5 was stationed at NAS Oceana, at Norfolk, Virginia. During March of 1943 the unit was redesignated as VS-37. The Kingfisher was used to scout the coast line and was often manned by Civil Air Patrol or U.S. Coast Guard air crews. (Naval History Center)

The pilot of this OS2N-1 Kingfisher holds the gas hose as the gunner/observer fills the wing tanks. This aircraft has obviously seen quite a bit of action judging by the patches on the cowl ring. Fuel is being provided by the USS CASCO (AVP-12) in Massacre Bay Attu, Alaska, during July of 1943. (National Archives)

Aviation Machinists Mates are giving an OS2N-1 Kingfisher of VS-37 a good cleaning on the ramp at NAS Norfolk, Virginia. This cleaning was necessary to prevent salt water corrosion. The aircraft in the background are Grumman TBF Avenger torpedo bombers. (National Archives)

Ground crewmen pull the prop through on an OS2N-1 in Alaska as they try to start the cold Pratt and Whitney R-985. The engine was usually started by a Type B (17 gram) cartridge, but in cold weather it was necessary to turn the engine over by hand until some oil could be run through the cold engine. (National Archives)

Destroyer Operations

Both the OS2N-1 and OS2U-3 were used to test the use of scout/observation aircraft on destroyers. Conceived as an idea to extend the eyes of the fleet even further by making more scout aircraft available not just from the battleships and cruisers, but also from the much smaller destroyers, six FLETCHER Class destroyers, USS PRINGLE (DD477), USS STANLEY (DD478), USS HUTCHINS (DD476), USS STEVENS (DD479), USS HALFORD (DD480) and USS LEUTZE (DD481) were chosen to be modified with a scout plane catapult and the associated aircraft handling and refueling facilities. Although chosen for catapult installation, the LEUTZE (DD481) never received the equipment.

The modification process took place during late 1940, after a successful test by the destroyer USS NOA (DD343) with the Curtiss XSOC-1 during early 1940. The catapult replaced the number three gun turret and the torpedo tubes. This reduced the fire power of the destroyers, but increased observation range for the fleet, since the destroyers normally ranged out ahead of the main force.

Space was limited on the destroyer and maintenance was a constant problem. The aircraft had to be launched from the starboard side, since the catapult could not rotate a full 360 degrees. Recovery was made to port since the lifting crane was on that side of the ship and recovery had to be made with the ship at almost a standstill. In peacetime this would not be a problem, but in combat, it put both the aircraft and ship in grave danger, especially from submarine attack.

These and other related problems prompted the Navy to remove the catapult and associated equipment from the HUTCHINS, PRINGLE and STANLEY during 1943. Although the STEVENS and HALFORD had experienced no operational problems, their equipment was removed during early 1944. These two ships, however, did use their aircraft in combat at Marcus Island. The HALFORD also took part in the Wake Island raid and the STEVENS saw action in the Tarawa raid of December 1943.

The OS2N-1s assigned to destroyer operations were: BuNo 01432 to the STEVENS, BuNo 01505 to the HUTCHINS and BuNo 01432 to the STANLEY. BuNo 5315, an OS2U-3, was attached to the PRINGLE. The aviation department aboard each consisted of the aircrew plus an aviation machinist mate and an aviation ordnanceman.

An armorer loads a 100 pound bomb on an OS2N-1 aboard the destroyer USS STEVENS (DD479) just prior to the invasion of Tarawa Island in December of 1943. THe Kingfisher had a very short career aboard destroyers because their small size led to less than adequate facilities for float plane operations. (National Archives)

USS STANLEY (DD478) was one of five FLETCHER class destroyers modified with catapults for float plane operations between 1942 to 1944. The catapult and handling equipment was installed in the area normally occupied by the number three 5 inch gun turret and aft torpedo tubes. Problems with flight operations from the small ships led the Navy to abandon the concept by early 1944. (National Archives)

XSO2U-1

During 1938, the Navy began searching for a replacement for the Curtiss SOC to serve aboard cruisers. The new aircraft would have to possess a greater range and top speed, carry more radio equipment and be able to handle bombs and depth charges, but still be able to fit on all cruiser class ships. To meet this requirement, the aircraft would need folding wings for storage purposes.

The specification also called for the aircraft to be powered by a 450 hp Ranger inverted Vee 12 cylinder air cooled engine. Vought-Sikorsky responded to the specification with their design No. 403. The aircraft resembled an XOS2U-1, but with a higher mid-wing and squared off cowling to house the XV-770-4 Ranger engine. The prototype was built under contract number 61265 issued on 21 June 1938. Designated the XSO2U-1 the prototype first flew on wheeled landing gear during July of 1939. During December of that same year it made its first flight on floats. The XSO2U-1, like the XOS2U-1, had interchangeable wheeled/float landing gear for versatility.

Curtiss also presented an aircraft to the Navy, the XSO3C-1. This aircraft was very similar to the XSO2U-1. Tests made by the Navy showed the Vought design to be superior in all respects; however, the contract for 300 aircraft went to Curtiss since Vought was fully involved with production of the OS2U-1 Kingfisher and the F4U Corsair.

The XSO2U-1 had a wing span of thirty-eight feet two inches, a length of thirty-six feet one inch, a height of fifteen feet eleven inches and wing area of 300 square feet. The fuselage and wings employed the spot welded construction technique pioneered on the OS2U, while all moveable control surfaces were fabric covered. Empty and maximum loaded weight was 4,016 and 5,624 pounds respectively. Total fuel capacity was 128 gallons carried in a fuselage tank.

The 450 hp XV-770-4 Ranger engine gave the XSO2U-1 a top speed of 190 mph at 9,000 feet and a landing speed of 55 mph. During the service life of the XSO2U-1 prototype, two different Ranger engines were installed. The XV-770-4 had the oil cooler intake on the front of the engine cowling and the XV-770-6 had the oil cooler intake mounted on the port side of the cowling. Cooling problems plagued the engine throughout its short service life on the Curtiss SO3C. The only reason that airframe manufacturers even considered using the XV-770 was due to its availability during the wartime period. The Ranger XV-770 engine was also used unsuccessfully in several other aircraft, notably the Beech XAT-14 and the Bell XP-77.

The XSO2U-1 was stressed for dive bombing and could carry either bombs or depth charges on a single pylon under each wing (similar to the OS2U-1). The XSO2U-1 was armed with a fixed .30 caliber M2 Colt Browning machine gun mounted on the starboard side of the engine synchronized to fire through the propeller arc and a flexible .30 caliber M2 gun in the rear cockpit for the radio operator/observer.

Early flight tests on floats indicated the aircraft suffered from some directional control problems, but this was solved with the addition of a stabilizing strake that ran under the fuselage from the rear of the float to just under the tail. The wings folded in much the same manner as the Grumman TBF Avenger, rotating 90 degrees and laying flat against the fuselage sides.

The Curtiss SO3C Seagull proved to be a dismal failure, not only for the Navy, but also for the British who were supposed to receive over 320 aircraft. After an initial batch of 250 had been delivered during 1942, the British refused to accept any more of the Seamens as they were called in Fleet Air Arm service (British crews gave the aircraft the more appropriate name of Sea Cow).

The XSO2U-1 was used as a test bed by Ranger Engine Corporation from July of 1942 until July of 1944 assisting in design studies for the Bell XP-77, a small interceptor and the Edo XOSE-1, a single seat float plane ordered by the Navy on 11 January 1944. When these tests were completed, the XSO2U-1 was stricken from the Navy inventory on 6 July 1944.

A pair of sailors help steady the SOX2U-1 prototype as it is prepared for sea trials on Long Island Sound during December of 1939. The 450 hp Ranger engine proved to be prone to overheating and the aircraft was cancelled during late 1942.